IMPROVING TRANSPORTATION TO FIGHT CLIMATE CHANGE

by Heather C. Hudak

FOCUS READERS
NAVIGATOR

WWW.FOCUSREADERS.COM

Focus Readers is distributed by North Star Editions:
sales@northstareditions.com | 888-417-0195

Produced for Focus Readers by Red Line Editorial.

Content Consultant: Don MacKenzie, PhD, Associate Professor of Civil & Environmental Engineering, University of Washington

Photographs ©: Shutterstock Images, cover, 1, 4–5, 7, 8–9, 10, 14–15, 16, 19, 21, 22–23, 25, 27, 29; Red Line Editorial, 13

Library of Congress Cataloging-in-Publication Data
Library of Congress Cataloging-in-Publication Data is available on the Library of Congress website.

ISBN
978-1-63739-273-7 (hardcover)
978-1-63739-325-3 (paperback)
978-1-63739-424-3 (ebook pdf)
978-1-63739-377-2 (hosted ebook)

Printed in the United States of America
Mankato, MN
082022

ABOUT THE AUTHOR

Heather C. Hudak lives in the foothills of the Rocky Mountains. She has written hundreds of kids' books on all kinds of topics. When she is not writing, Hudak loves to travel. She hopes to fly around the world in a hydrogen-powered airplane one day.

TABLE OF CONTENTS

CHAPTER 1

CLEANING UP THE SKIES

In December 2021, a plane took off from Chicago, Illinois. Two hours later, it landed in Washington, DC. To the passengers, the flight felt normal. But that meant it was a huge success. The plane had used a different type of fuel. One engine had used 100 percent sustainable aviation fuel (SAF).

A flight from Chicago, Illinois, to Washington, DC, can take more than 1,000 gallons (3,785 L) of fuel.

Most planes run on regular jet fuel. That fuel is made from fossil fuels. Fossil fuels come from long-dead plants and animals. Burning them releases carbon dioxide (CO_2) into the atmosphere. CO_2 is a greenhouse gas. Greenhouse gases trap heat in the atmosphere. They are causing **climate change**.

Burning SAF also releases CO_2. But this fuel comes from cooking oil, grease, and farm waste. These sources come from plants. Plants take in CO_2 when they grow. As a result, SAF produces fewer overall CO_2 **emissions** than fossil fuels. For this reason, SAF can help in the fight against the climate crisis.

In 2020, air travel produced 2.5 percent of CO_2 emissions. Scientists expected this number to rise.

The 2021 flight was the first to use only SAF in an engine. But the fuel still had a long way to go. As of 2021, SAF made up less than 0.1 percent of all jet fuel. Even so, it was one of many helpful tools.

CHAPTER 2

TRANSPORTATION HISTORY

For most of history, people relied on very basic forms of energy. They used muscles or animals to move things. They burned wood as fuel for heat or cooking.

That changed in the 1700s. Scientists built the first steam engines. Burning coal produced the steam for these engines. That steam held a lot of energy. Steam

Trains were some of the first widespread methods of transportation to use steam engines.

In 2020, more than 270 million vehicles were on US roads.

engines soon began powering trains and ships. It was the start of the **Industrial Revolution**.

In the mid-1800s, scientists developed the first internal combustion engines. These engines burn air and fuel to produce power. They used oil instead of coal. Internal combustion engines led to even more kinds of transportation.

AIR POLLUTION

Burning fossil fuels to travel doesn't just change the climate. It also creates air pollution. Cars, trucks, and other vehicles produce smog and soot. People breathe in this pollution. It can cause health problems such as asthma. In fact, air pollution causes millions of deaths each year. It does not harm everyone equally. People in low-income countries suffer the most. In the United States, low-income and Black communities are most harmed.

Factories produced huge numbers of cars. Then inventors developed airplanes in the early 1900s. Roads and railways stretched across entire continents. Cargo ships carried goods across the oceans.

By the end of the 1900s, fossil fuels powered most forms of transportation. Burning these fuels released huge amounts of CO_2 into the air. All that CO_2 is leading to climate change.

By the 2020s, transportation created approximately 20 percent of the world's CO_2 emissions. And people experienced the climate crisis all over the world. Sea levels rose. Droughts became longer. Wildfires became more severe. Extreme temperatures were more common, and storms were stronger. People needed to take action to stop the crisis from getting worse.

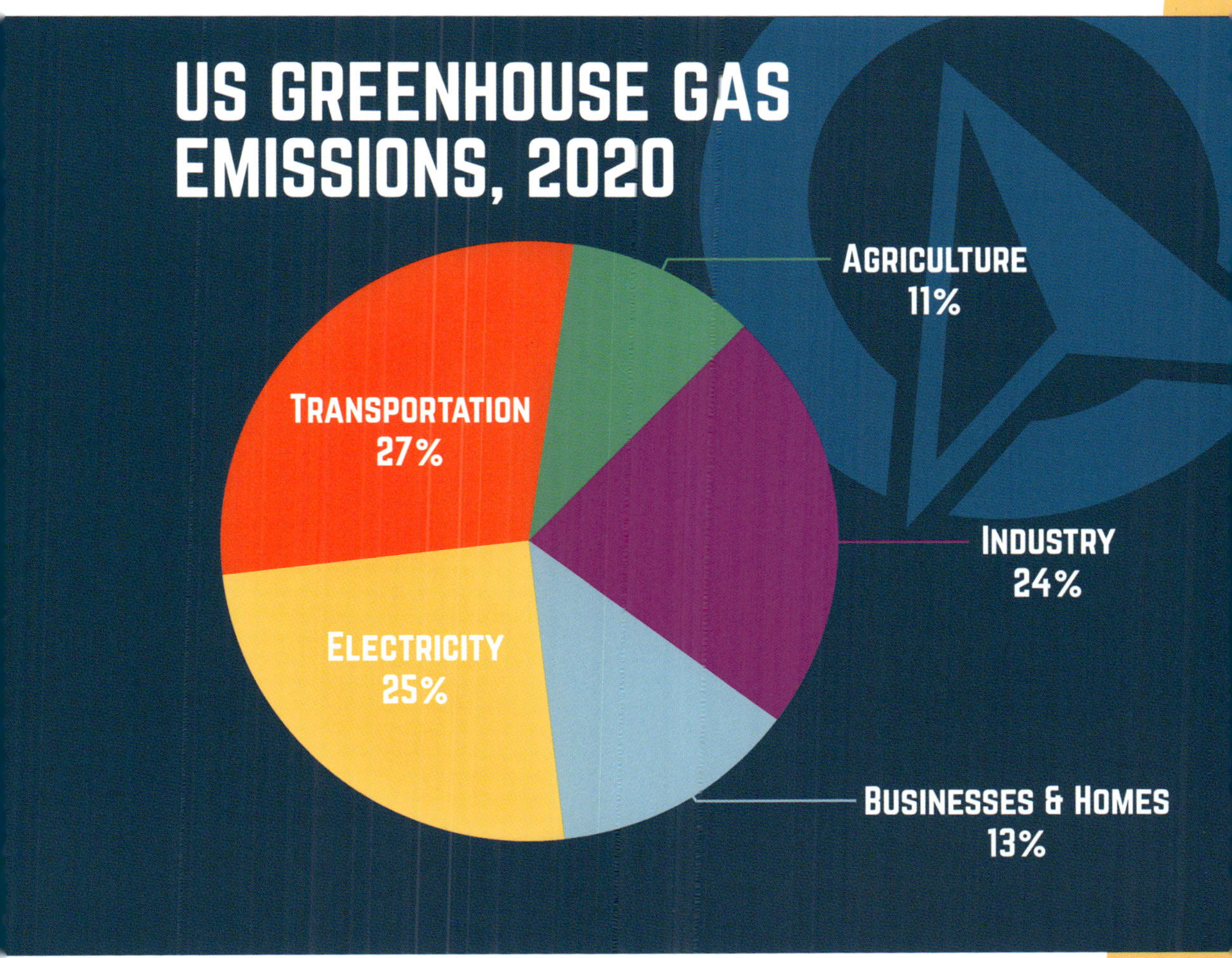

CHAPTER 3

GREENING SPEED

Scientists have improved transportation in many ways. For example, electric vehicles (EVs) do not use fossil fuels. Instead, electricity powers them. EVs have one or more electric motors. Batteries power the motors. Then people plug their EVs into a power source. That recharges the batteries.

Many people charge their electric cars in their garages at home.

can only carry a few people in a single trip. Full buses and trains carry far more people. They can produce much lower emissions per person than cars do.

In addition, many places have high-speed rail. These trains run on a

MAGLEV TRAINS

Some of the world's fastest trains are magnetic levitation trains. They are called maglev trains for short. They can travel more than 250 miles per hour (402 km/h). A magnetic coil runs along the track. The coil pushes against magnets on the bottom of the train. This makes the train float. Then electricity runs through the coil. It creates a **magnetic field**. The magnetic field pulls the train forward and pushes it from behind.

Maglevs do not touch the tracks, so there is less wear on the trains. This means they last longer.

track just like regular trains. But the trains travel much faster than cars. Some rail systems go between cities and countries. People can use trains instead of planes. That cuts down on emissions. Shifting to these forms of transportation can help slow climate change.

HYPERLOOPS

In the early 2020s, a few companies were working on hyperloops. A hyperloop is an ultra-high-speed railway. Riders get inside a pod. It travels inside a tube. Similar to a maglev, the pod floats on magnets. Floating gets rid of friction. Friction is the resistance of surfaces against one another. It slows objects down. Plus, vacuum pumps suck the air out of the tube. Air also slows objects down. Without air or friction, the pod can go faster.

Hyperloops would use very little energy to run. That could make them a cleaner choice for travel. However, a hyperloop would take a lot of time and money to build. It requires large amounts of land, too. As of 2022, none had been built yet.

Some hyperloop designs aim for speeds of more than 600 miles per hour (966 km/h).

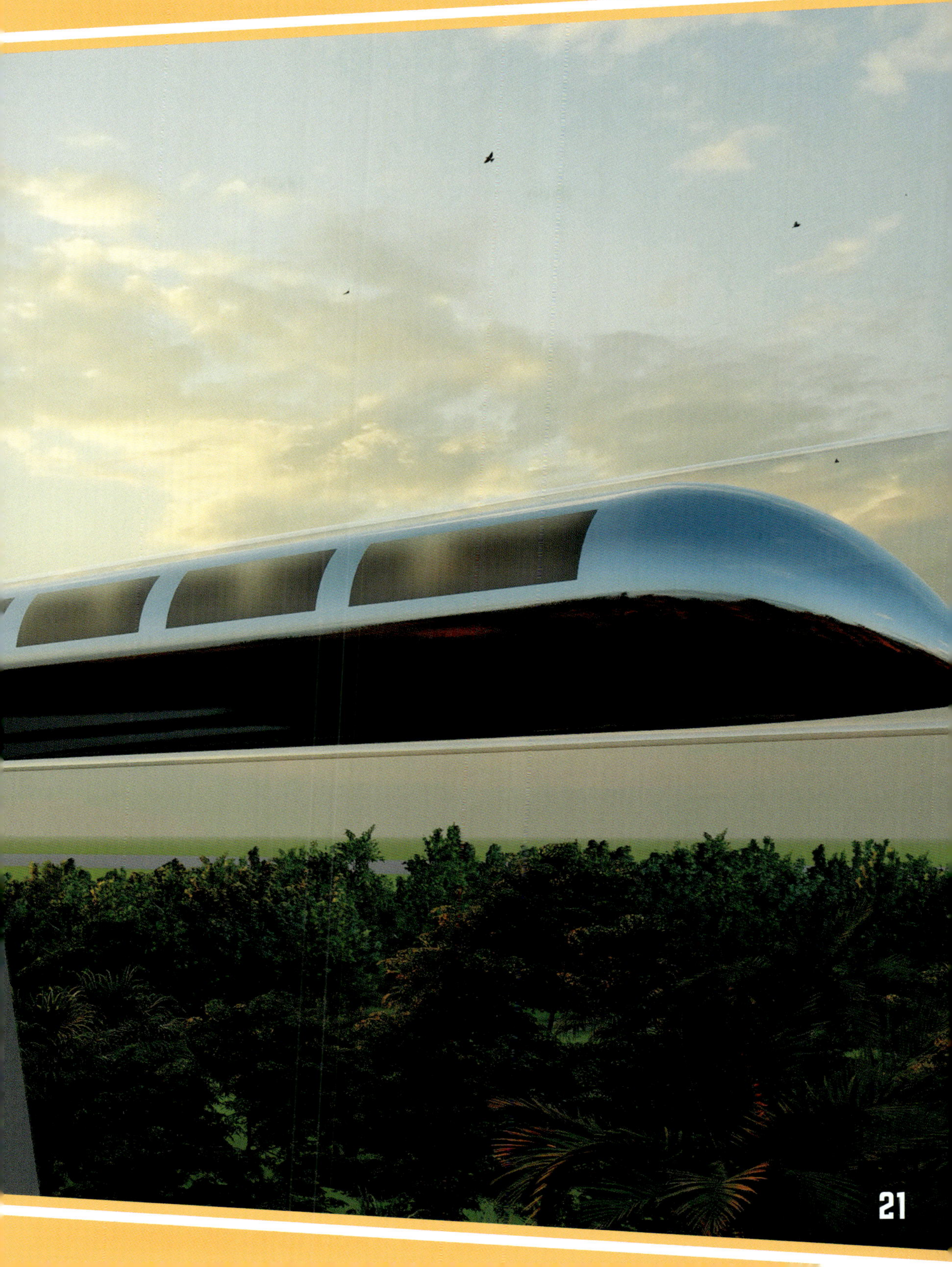

CHAPTER 4

ADDRESSING ROADBLOCKS

Changing the way people get around is not easy. For example, an increase in demand for EVs could lead to challenges. EV batteries often use lithium and cobalt. Mining these metals can be costly. It can harm miners and the environment. In response, scientists are working on ways to recycle EV batteries. They are also

Salt flats, such as Chile's Salar de Atacama, are common sources of lithium.

developing new batteries. The batteries would use less lithium and cobalt. These shifts would reduce the need to mine.

Another challenge involves **electrical grids**. Switching to EVs might mean using less energy overall. But it would mean using more electricity. Areas would need to improve their electrical grids. That way, they could handle the extra demand.

EVs also require new **infrastructure**. In the early 2020s, gasoline stations were more common than charging stations. Plus, people could refuel gasoline-powered cars in minutes. The fastest charging stations took half an hour. Others took several hours. As

In 2022, the United States had more than three times as many gasoline stations as charging stations.

a result, it was still easier to drive a gasoline-powered car for long trips.

Scientists were also working on better batteries for large EVs. Electric trucks were nearly ready by the early 2020s. And several cities were using electric buses. But electric ships and planes still had a long way to go.

For this reason, scientists were also developing other ideas. Some ideas used older technologies. For instance, airships are similar to blimps. They use gases that are lighter than air, such as helium. To lift

CITY PLANNING

Improving transportation often involves changing vehicles. But improving cities also matters. For example, many US cities are built around cars. Cities are filled with highways, wide roads, and parking lots. These features make it hard for people to walk or bike. They make building train systems difficult, too. However, many cities and towns are making changes. They are turning car lanes into bike paths and bus routes. These kinds of shifts help cut greenhouse gas emissions.

In the early 2020s, airships often were used to advertise companies.

off, airships release their heavy air. The helium left inside is lighter than the air outside. Then the aircraft floats.

Airships use combustion engines to push them along. These engines produce some emissions. But airships do not need to burn fuel to stay afloat. So, they produce far fewer greenhouse gas emissions than regular aircraft.

People have used wind power to sail across the seas for centuries. Wind is a clean energy resource. Scientists worked on wind-powered ships for today's world. One company designed wind-powered cargo ships. The designs had huge sails. The sails could turn in all directions. That ability would help them capture as much wind as possible. Many wind-powered ships do not burn any fossil fuels.

By the early 2020s, people everywhere could feel the climate crisis. Scientists agreed that massive changes were necessary to avoid the most severe

In 2019, activist Greta Thunberg sailed across the ocean instead of flying to raise awareness about climate change.

effects. And they agreed that those shifts needed to happen quickly. One major area for change was transportation. Thankfully, much of the technology was ready. People now needed to work together to make the technology a reality.

FOCUS ON
IMPROVING TRANSPORTATION

Write your answers on a separate piece of paper.

1. Write a sentence describing the main ideas of Chapter 4.
2. What clean transportation would you like to use? Why?
3. In the early 2020s, what percent of the world's greenhouse gas emissions came from transportation?

    A. less than 5 percent
    B. approximately 20 percent
    C. more than 60 percent

4. What is one reason electric vehicles can cut down on greenhouse gas emissions?

    A. Electric vehicles are less efficient than gasoline-powered vehicles.
    B. Electric vehicles are as efficient as gasoline-powered vehicles.
    C. Electric vehicles are more efficient than gasoline-powered vehicles.

Answer key on page 32.

GLOSSARY

climate change
A human-caused global crisis involving long-term changes in Earth's temperature and weather patterns.

efficient
Accomplishing as much as possible with as little effort or as few resources as possible.

electrical grids
Complex networks that deliver electricity from power sources to users.

emissions
Chemicals or substances that are released into the air, especially ones that harm the environment.

Industrial Revolution
Starting in Great Britain in the 1700s, a huge economic shift involving the use of powerful machines and mass production.

infrastructure
The systems, such as roads, water supplies, and energy distribution, that a city, state, or nation needs to function.

magnetic field
The space around an object in which its magnetic force can be detected.

TO LEARN MORE

BOOKS

Conley, Kate. *Designing City Transport*. Minneapolis: Abdo Publishing, 2019.

Heitkamp, Kristina Lyn. *Electric Vehicles*. Lake Elmo, MN: Focus Readers, 2022.

Klepeis, Alicia Z. *The Future of Transportation: From Electric Cars to Jet Packs*. North Mankato, MN: Capstone Press, 2020.

NOTE TO EDUCATORS

Visit **www.focusreaders.com** to find lesson plans, activities, links, and other resources related to this title.

INDEX

Answer Key: 1. Answers will vary; **2.** Answers will vary; **3.** B; **4.** C